SCIENCE QUESTIONS

WHY DO VOLCANOES ERUPT?

by Elizabeth Andrews

Cody Koala
An Imprint of Pop!
popbooksonline.com

abdobooks.com
Published by Pop!, a division of ABDO, PO Box 398166, Minneapolis, Minnesota 55439.

Printed in the United States of America, North Mankato, Minnesota

102021
012022

THIS BOOK CONTAINS RECYCLED MATERIALS

Cover Photo: Shutterstock Images
Interior Photos: Shutterstock Images, 1, 5 (top, bottom left, bottom middle), 7, 8, 11, 13, 14, 16, 19, 20–21

Editor: Grace Hansen
Series Designer: Laura Graphenteen

Library of Congress Control Number: 2021942298

Publisher's Cataloging-in-Publication Data
Names: Andrews, Elizabeth, author.
Title: Why do volcanoes erupt? / by Elizabeth Andrews
Description: Minneapolis, Minnesota : Pop!, 2022 | Series: Science questions | Includes online resources and Index.
Identifiers: ISBN 9781098241117 (lib. bdg.) | ISBN 9781098241810 (ebook)
Subjects: LCSH: Volcanoes--Juvenile literature. | Volcanic eruptions--Juvenile literature. | Magmas--Juvenile literature. | Children's questions and answers--Juvenile literature.
Classification: DDC 500--dc23

Hello! My name is

Cody Koala

Pop open this book and you'll find QR codes like this one, loaded with information, so you can learn even more!

Scan this code* and others like it while you read, or visit the website below to make this book pop.

popbooksonline.com/volcanoes-erupt

*Scanning QR codes requires a web-enabled smart device with a QR code reader app and a camera.

Table of Contents

Chapter 1

A Force of Nature

It can make the ground shake. It can make fire rain from the sky. It can create new landforms. It's a volcanic **eruption**!

Watch a video here!

Chapter 2

Earth's Layers

Planet Earth has four layers. The crust is the rocky surface. Below that is the mantle. It is made of solid and **molten** rock.

Learn more here!

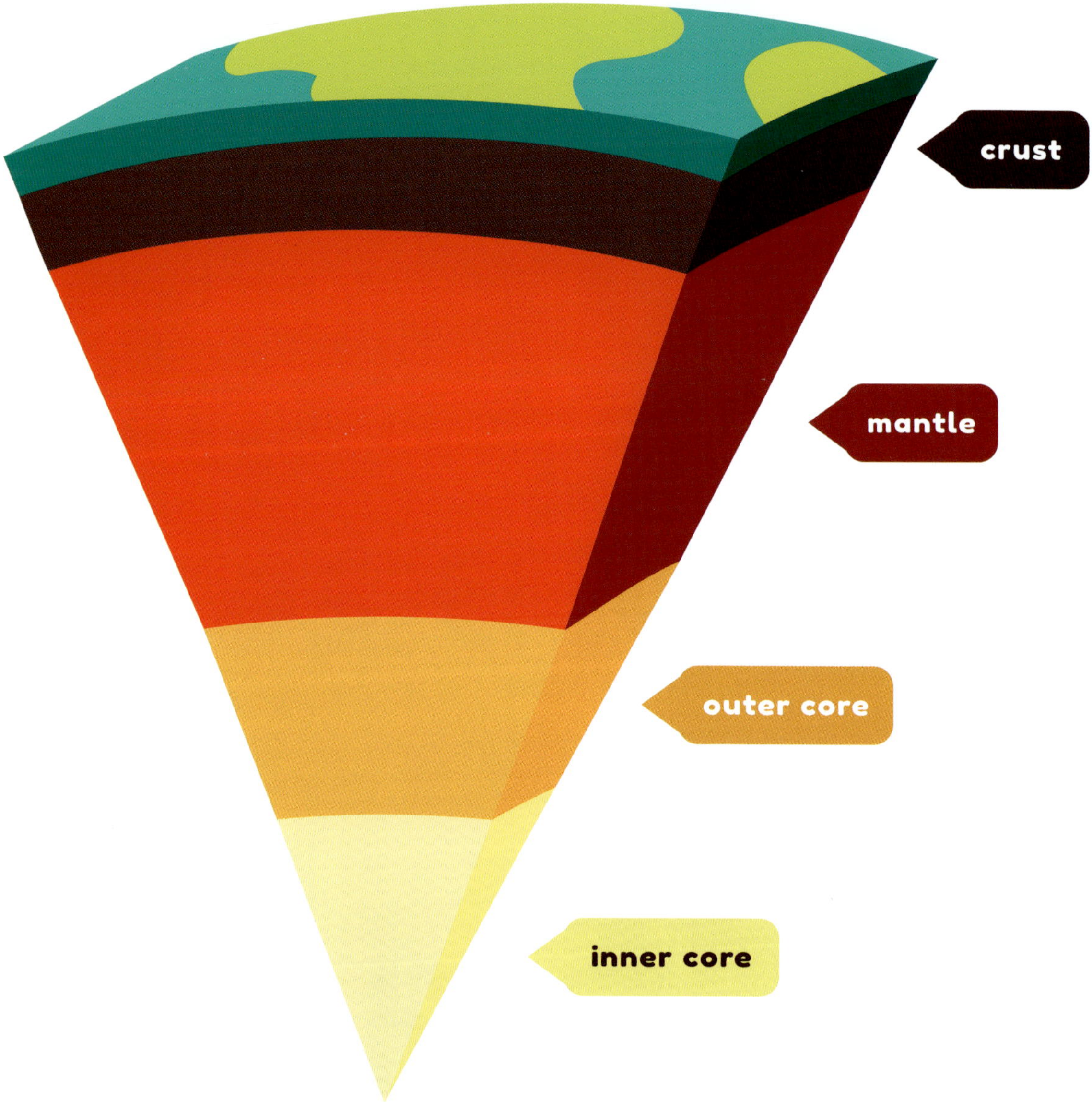
crust
mantle
outer core
inner core

Earth's **core** has an inner and an outer layer. Both are made of metals. The outer core is liquid metal. The inner core is solid metal.

The crust of Earth is split into plates of rock. They fit together like a puzzle. The plates are floating on top of the mantle’s molten rock. Many volcanoes form along the edges of the plates.

When two plates meet it is called a plate boundary. You can walk along one in Iceland.

Chapter 3

What Is a Volcano?

Volcanoes are **vents** on Earth's crust. Many of them look like mountains. The opening of the volcano is called the crater. Below the crust of the volcano is a main vent.

Learn more here!

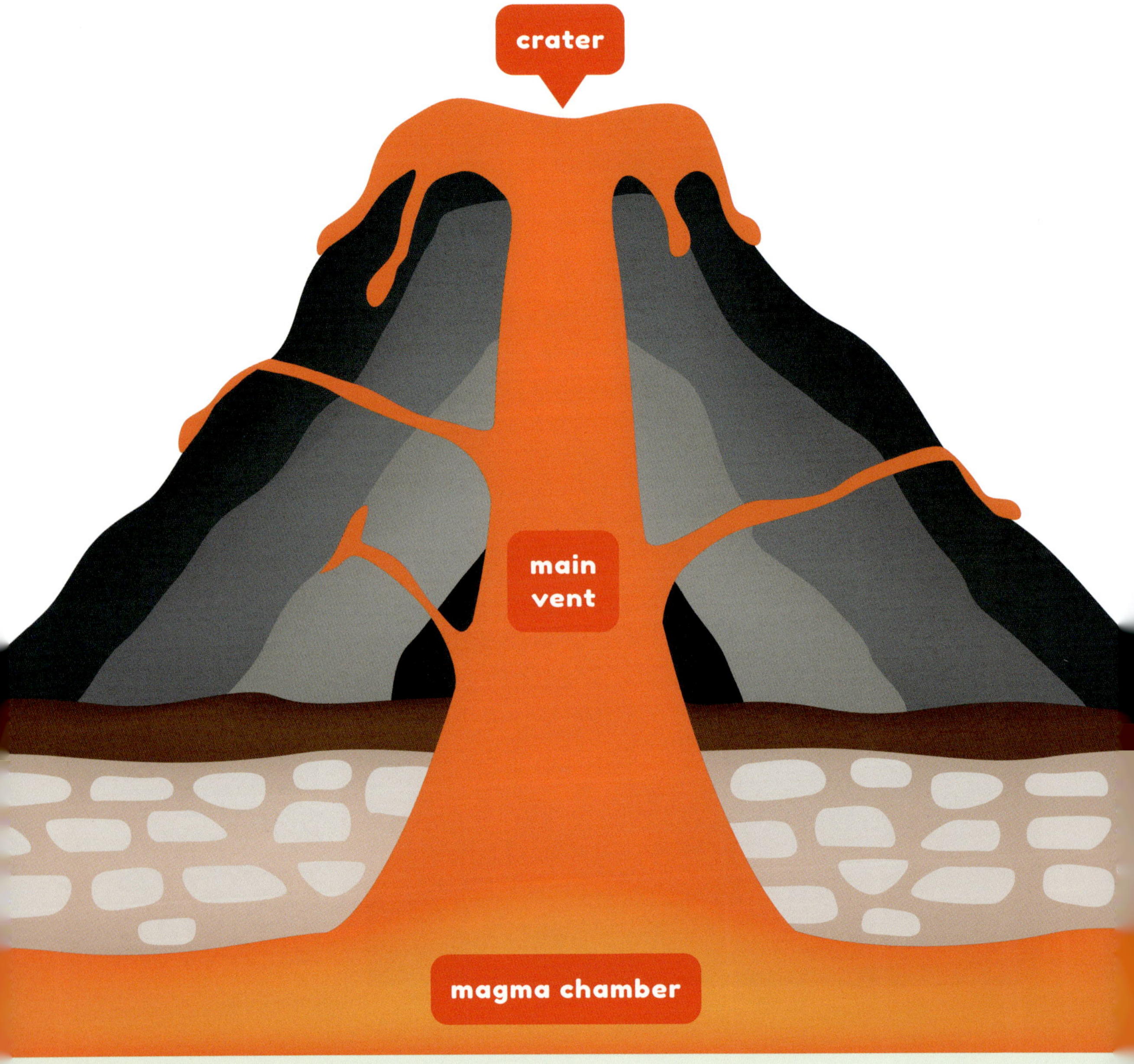
crater
main vent
magma chamber

The main vent connects a magma chamber to an opening on the crust. Magma is very hot, liquid rock below the crust. It comes from the mantle. The magma chamber is a large pool of magma that rises to the surface.

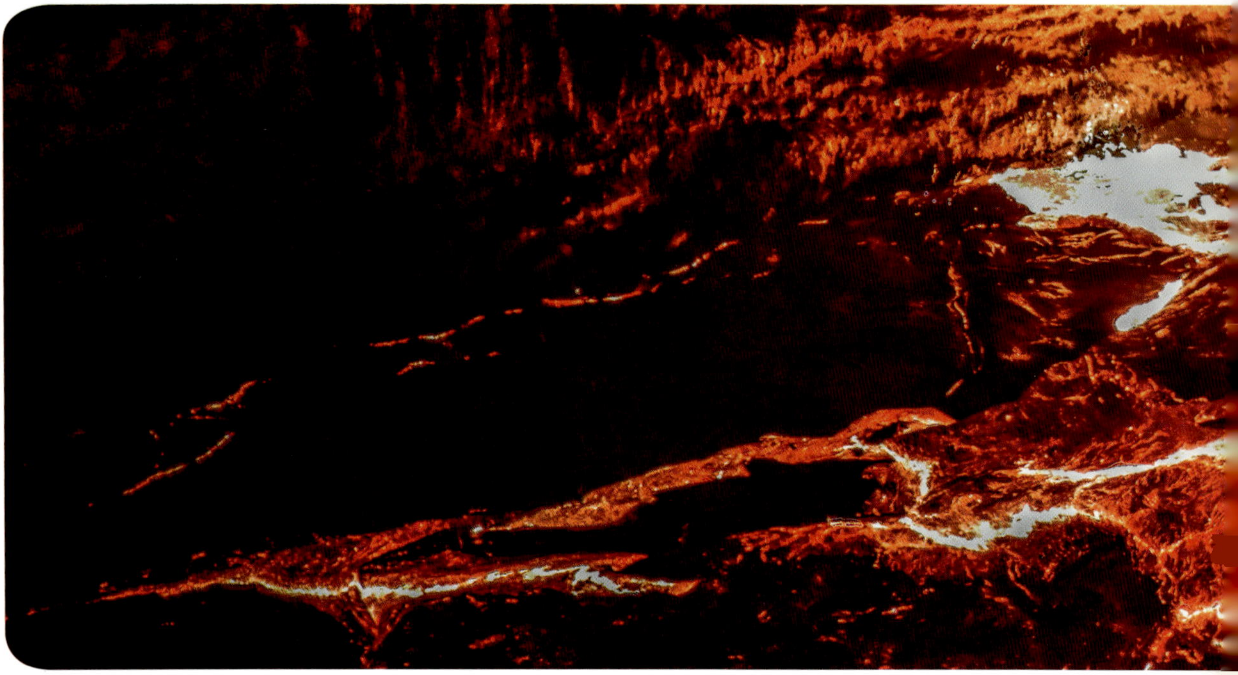

Magma can rise because it is less **dense** than solid rock. It flows through cracks in the rocks of the crust to

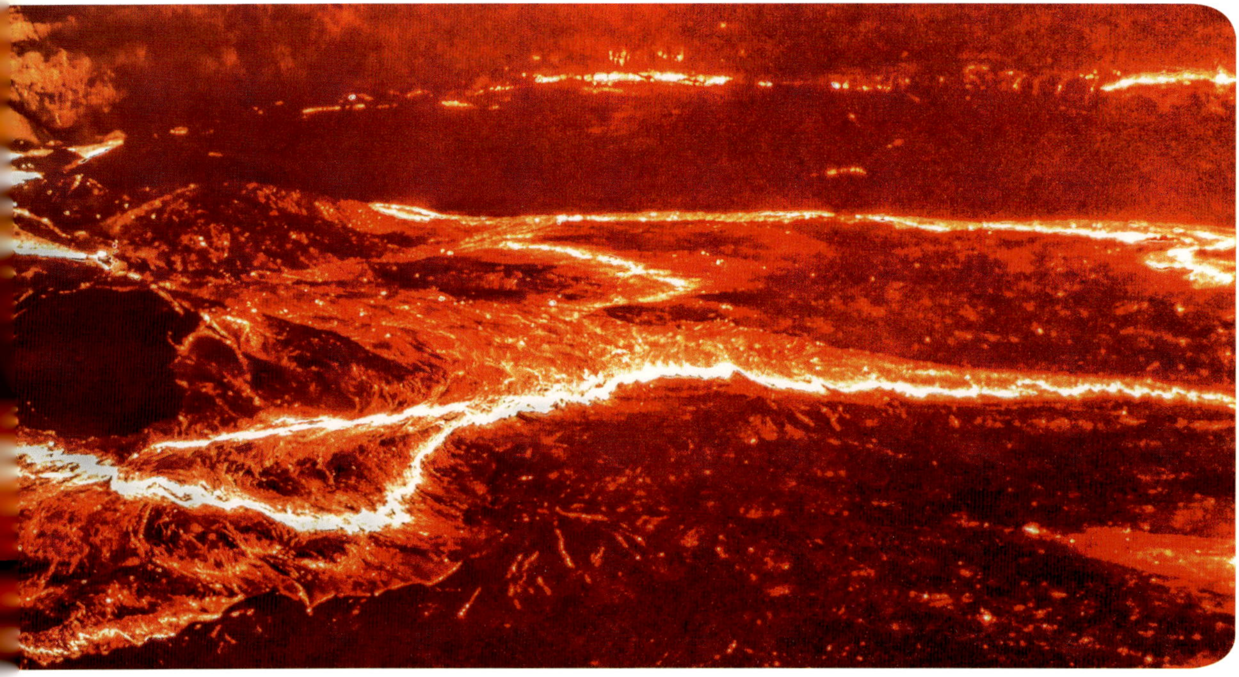

collect in the chamber. The chamber also holds gas created by the magma.

Chapter 4

Eruption!

As magma rises, the **pressure** in the chamber builds. All the gas and **molten** rock filling the space needs a place to go. It will be pushed up and out through the **vent**. This is an **eruption**.

When magma reaches the surface it is called lava.

Complete an activity here!

Eruptions usually last for two months. Sometimes the eruption is explosive. Gas, lava, and pieces of rock will

shoot into the air. An eruption can also be calm. Lava might flow gently from the volcano's crater.

Making Connections

Text-to-Self

Would you ever want to see a volcano erupt? How would you stay safe while watching?

Text-to-Text

Have you read any other books about natural disasters? How were they similar to and different from a volcanic eruption?

Text-to-World

Do you think it's a good idea for people to live near volcanoes? Explain why or why not.

Glossary

core – the very hot, dense center of Earth.

dense – tightly packed.

erupt – to force out or release, sometimes suddenly.

molten – made liquid by very high heat.

pressure – a steady force upon a surface.

vent – an opening that allows gas or liquid to pass through.

Index

Online Resources

popbooksonline.com

Thanks for reading this Cody Koala book!

Scan this code* and others like it in this book, or visit the website below to make this book pop!

popbooksonline.com/volcanoes-erupt

*Scanning QR codes requires a web-enabled smart device with a QR code reader app and a camera.